The Convention on Biological Diversity and Product Commercialisation in Development Assistance Projects: a Case Study of LUBILOSA

CABI *Bioscience* is a division of **CAB** *International*, an inter-governmental, not-for-profit, mission-oriented organisation dedicated to improving human welfare world-wide through the dissemination, application and generation of scientific knowledge in support of sustainable development. Emphasis is placed on agriculture, forestry, human health and the management of natural resources and particular attention is given to the needs of developing countries.

CABI *Bioscience*'s **Biopesticides Programme** is committed to the development and use of biopesticides as safe, environmentally friendly alternatives to chemical pesticides. The Programme carries out collaborative inter-disciplinary research and development, offers training in insect pathology, runs the International Biopesticide Consortium for Development (IBCD), disseminates information and promotes the role and value of biopesticides in sustainable crop production, poverty alleviation and wealth generation.

Biopesticides series

1. *Chemical Pesticide Markets, Health Risks and Residues*
   J. Harris

2. *Priorities in Biopesticide Research and Development in Developing Countries*
   J. Harris and D.R. Dent

3. *The Convention on Biological Diversity and Product Commercialisation in Development Assistance Projects: a Case Study of LUBILOSA*
   D.R. Dent and C. Lomer

Biopesticides are biological pesticides based on beneficial insect and weed pathogens and entomopathogenic nematodes. Pathogens used as biopesticides include fungi, bacteria, viruses and protozoa. Produced, formulated and applied in appropriate ways, such biopesticides can provide ecological and effective solutions to pest problems.

The aims of the Biopesticides series are to more widely appraise and promote the role and value of biopesticides as alternatives to chemical pesticides and to improve awareness of the opportunities offered by biopesticides.

The series has been developed by the Biopesticides Programme at CABI *Bioscience* as part of its mission to disseminate information and promote the role and value of biopesticides.

The Convention on Biological Diversity and Product Commercialisation in Development Assistance Projects: a Case Study of LUBILOSA

Biopesticides Series No. 3

David R. Dent

CABI Bioscience (UK Centre)
Ascot, UK

Chris Lomer

Royal Veterinary and Agricultural University (KVL)
Institute for Ecology, Copenhagen, Denmark
(Formerly) International Institute of Tropical Agriculture
Cotonou, Benin

CABI *Publishing*

CABI *Publishing* is a division of CAB *International*

CABI Publishing
CAB International
Wallingford
Oxon OX10 8DE
UK

Tel: + 44 (0)1491 832111
Fax: +44 (0)1491 833508
Email: cabi@cabi.org
Web site: www.cabi.org

CABI Publishing
10 E 40th Street
Suite 3203
New York, NY 10016
USA

Tel: + 1 212 481 4018
Fax: + 1 212 686 7993
Email: cabi-nao@cabi.org

A catalogue record for this book is available from the British Library, London, UK.
A catalogue record for this book is available from the Library of Congress, Washington DC, USA.

ISBN 0 85199 577 2

Printed and bound in the UK by Biddles Ltd, Guildford and King's Lynn, from copy supplied by the authors.

Contents

Acronyms vii

Acknowledgements ix
    Disclaimer ix

Executive Summary...xi

Introduction 1

The LUBILOSA Programme — History...3
    The Convention on Biological Diversity and the LUBILOSA
        Programme 4
    Origin, distribution and access to the LUBILOSA isolate IMI
        330189 5

Product Development 7

Green Muscle®: the Product 9
    Taxonomy 9
    Active material..9
    Formulations 9
    Storage characteristics...9
    Application..10
    Status of registration 10

Green Muscle®: the Route to Dissemination and Technology Transfer 11

Disclosure, Sharing and Exchange of Information................................13
    Public sector scientists in general..13

Research providers 13
Donors supporting R&D projects 14
Commercial companies 14
Dealing with confidentiality at project level 14
Ownership of the intellectual property 17

Commercial Company Collaboration 21
Green Muscle®: positive attributes 22
Green Muscle®: negative attributes 22
Approaches to commercial companies 24
Commercial company interest in biopesticides 25
Which companies? 25
Companies with access to the acridid market 26
The basis of LUBILOSA collaboration with commercial
      companies 27
Sharing of benefits arising from LUBILOSA 28

Conclusions 33

Recommendations 35

References 37

Annex A: Model Confidentiality Agreement 39

Annex B: Multi-Sponsor Intellectual Property Rights Agreement 43

Annex C: CGIAR Material Transfer Agreement 49

Acronyms

BCP	Biological Control Products (commercial producer in Southern Africa)
CBD	Convention on Biological Diversity
CGIAR	Consultative Group on International Agricultural Research
CIDA	Canadian International Development Agency
CILSS	Comité Inter-états pour la Lutte contre la Secheresse au Sahel
DFID	Department for International Development (UK)
DGIS	Directorate General for International Development (Netherlands)
DFPV	Departément de Formation en Protection des Vegetaux
FAO	Food and Agriculture Organization (of the United Nations)
GTZ	Deutsche Gesellschaft für Technische Zusammenarbeit
IITA	International Institute of Tropical Agriculture
IPM	Integrated Pest Management
IPR	Intellectual Property Rights
LUBILOSA	Lutte Biologique contre les Locustes et les Sauteriaux
MTA	Material Transfer Agreement
NGO	Non-government organisation
NPP	Natural Plant Protection (commercial producer for West Africa)
SDC	Swiss Development Co-operation
SME	Small to medium enterprise
USAID	United States Agency for International Development

Acknowledgements

Phase 3 of the LUBILOSA Programme was funded by the Canadian International Development Agency (CIDA), the Swiss Development Co-operation (SDC), the Directorate General for Development Co-operation of the Netherlands (DGIS) and the Department for International Development of the UK (DFID, formerly ODA).

The authors would like to thank the sponsors for their comments on the text and permission to publish extracts from their contracts held with CABI and for their time and contribution to this study. The authors would also like to thank all those in the LUBILOSA Programme who have contributed to the data and ideas presented herein.

Disclaimer

The views expressed in this publication are those of the authors, which do not necessarily correspond with those of the LUBILOSA sponsors or LUBILOSA Programme. Partial extracts or complete copies of contractual clauses reproduced herein are intended for information purposes only. You should seek legal advice before taking any action or making any decision based on this information. While all efforts have been made to ensure the accuracy and completeness of this information, neither the authors, LUBILOSA sponsors nor LUBILOSA Programme are responsible for any errors or omissions, or for the results obtained from the use of this information.

Executive Summary

The LUBILOSA Programme

The LUBILOSA Programme was initiated in 1989 and has been successful in developing a mycoinsecticide for the biological control of locusts and grasshoppers. Spores of the naturally occurring fungus *Metarhizium anisopliae* var. *acridum* (having a geographical range including Brazil, the Galapagos, Africa, Madagascar and Australia), suspended in an oil formulation, are sprayed to induce infections of target locusts and grasshoppers. The efficacy of the product, named Green Muscle®, has been demonstrated against all the major acridid species in Africa, through numerous field trials undertaken in collaboration with African National Programmes. The product provides an environmentally friendly alternative to chemical insecticides for the control of locusts and grasshoppers in Africa.

The Convention on Biological Diversity and the LUBILOSA Programme

The LUBILOSA Programme pre-dates the Convention on Biological Diversity (Rio de Janeiro 1992), however the principles and management of the Programme and the product Green Muscle® have been established in accordance with the benefit sharing and related provisions of the Convention on Biological Diversity.

- Full participation in research (Article 15.6);
- Equitable sharing of research results and benefits from commercial and other use (Article 15.7);
- Access to and transfer of technology (Article 16);
- Exchange of information (Article 17);
- Technical and scientific co-operation (Article 18);

- Participation in research and access to results and benefits of biotechnology (Articles 19.1 and 19.2);
- Financial resources (Article 20).

Research participation and collaboration

The LUBILOSA Programme trained 40 scientists and technicians from collaborating National Programmes in Africa in the techniques of mycoinsecticide development, production and application.

National Programmes have, through collaborative trials, contributed to the product's development and fully assessed and evaluated its potential for locust and grasshopper control in their own countries.

Equitable sharing of research results and exchange of information

A total of 120 scientific papers, conference proceedings and reports have been produced by LUBILOSA scientists and collaborators that detail the techniques, approaches and results of the research including, production, formulation, storage, ecotoxicology and application. These publications which are widely available provide sufficient information to enable anyone to develop, produce on an artisanal scale, formulate and apply a basic but highly effective mycoinsecticide for locust and grasshopper control.

All collaborators have been able to contribute to and receive copies of a regular Programme Newsletter disseminating information about the Programme and its achievements. Representatives of National Programmes have been invited to and attended numerous scientific meetings and Programme Management Meetings over the 10 year duration of the Programme.

Access to genetic resources

The isolate IMI 330189 of *Metarhizium anisopliae* var. *acridum* is the active ingredient of the mycoinsecticide. This isolate is maintained in culture by CABI *Bioscience* and is available on request for R&D and non-commercial purposes. The isolate can be requested for commercial purposes on the basis that at the point of commercialisation an appropriate agreement addressing the sharing of benefits is negotiated.

A patent has not been sought nor is there any intention to do so for the purposes of securing rights to the sole use of the isolate IMI 330189 as the active ingredient of a mycoinsecticide for locust and grasshopper control – any such action is not possible under the patent regulations of the UK and most other countries, with the exception of the USA.

Access to the technology

A UK patent for the oil formulation of Deuteromycete fungi (which includes *Metarhizium anisopliae* var. *acridum* isolate IMI 330189) was obtained on 15 March 1995 by CABI on behalf of the LUBILOSA Programme and its sponsors. The patent was obtained to ensure that other organisations/commercial companies would not apply for similar patents and thereby restrict access to the oil formulation technology, particularly in developing countries.

The licence by CABI, on behalf of LUBILOSA and its sponsors, of the formulated product to two commercial companies does not prevent or restrict other commercial or non-commercial use of the isolate IMI 330189 for biopesticide development and application. The licence restricts use of the technology of a sophisticated oil miscible flowable formulation that incorporates the isolate IMI 330189 only (known as Green Muscle®), the toxicological data associated with this and down stream processes that influence sophisticated industrial scale production. Developing country artisanal production and use of the isolate and the simple oil formulation technology is in no way restricted by these licences.

The licences issued to the companies include clauses that ensure the following conditions apply:

- good commercial practice with recourse to the appropriate transfer of public sector technology to the commercial sector;
- a reasonable price is charged for the sale of the product; the price must be competitive with other similar products to ensure general use and accessibility;
- there is a reasonable availability of the mycoinsecticide within sponsor core countries requiring such products;
- the monies generated from the commercial exploitation of the mycoinsecticide shall be credited to a Trust Fund and used in accordance with the declared objectives of the Fund.

Sharing of benefits arising from commercialisation of the product

The reason for licensing the production, marketing and sale of the mycoinsecticide to two commercial companies was to ensure that the product Green Muscle® could be produced on a sufficient scale at appropriate international standards of quality and performance for purchase and use, largely by international and national locust control programmes.

The royalties and licence fees generated from the sale of the product by the licencees will be accumulated in a Trust Fund for disbursement within Africa.

Introduction

The Convention on Biological Diversity that entered into force on 29 December 1993 addressed the crucial issues of conservation and sustainable use of biological diversity encompassing such matters as access to genetic resources, sharing benefits from the use of genetic material and access to technology. The development and commercialisation of products based on biological materials through development assistance projects need to incorporate the principles and practices defined in the CBD. This requires that projects address issues concerning research participation and sharing of results, access to and transfer of technology and the sharing of benefits arising from the R&D. Where projects involve just one or two partners the means by which the CBD can be implemented are more obvious. The situation is more difficult with a collaborative project that involves a large number of different types of partners (e.g. universities, national research institutes, extension agencies, NGOs, commercial companies), that are based in many countries, and which span many years of work.

The LUBILOSA Programme (see below) is an example of a programme implementing the CBD that involves a large number of partners from different types of organisation based in different geographical locations over a period of 10 years. Although the exploration and collection of isolates of *Metarhizium* in the LUBILOSA programme pre-dates the CBD, the spirit in which the collection was made is in-line with both the principles and details of the CBD. In addition, the CBD provides a clear articulation and legal basis for many of the issues faced by the LUBILOSA team in the implementation of their programme.

The LUBILOSA Programme — History

Locust outbreaks have been a problem to mankind for centuries. During a recent locust outbreak (1986–1989) the donor community committed US$275 million for locust and grasshopper control. Some 13 million litres of insecticide were applied, mostly malathion and fenitrothion. The very high cost of these control operations and the concerns expressed about their effectiveness and environmental impact led to unease among the international donor community. This unease led to the reassessment of locust control methods in a conference organised by the FAO entitled "Defining Future Research Priorities", held in October 1988. At this conference great interest was shown in the ideas put forward by the then International Institute of Biological Control — now CABI *Bioscience* — on the possibility of developing a biopesticide (based on naturally occurring fungi) for locust control (FAO, 1989). Within a year this interest had manifested itself into a collaborative programme of research and development entitled the "Biological Control of Locusts and Grasshoppers" (Greathead, 1992). This programme has since become known better by the acronym of its French translation "Lutte Biologique contre les Locustes et Sauteriaux" — that is LUBILOSA. The programme was funded by the Canadian International Development Agency (CIDA), the Directorate General for Development Co-operation of the Netherlands (DGIS), the Department for International Development of the UK (DFID, formerly ODA), Swiss Development Co-operation (SDC) and the United States Agency for International Development (USAID). It began in 1989 as a collaborative 3 year programme being undertaken by CABI *Bioscience*, the International Institute of Tropical Agriculture (IITA) and the Departément de Formation en Protection des Vegetaux (DFPV), Niger.

Since the beginning of LUBILOSA in 1989, the programme has moved through the whole process of development for a mycoinsecticide. It achieved this through three phases of the programme, each of 3 years duration.

Phase 1 (1989–1991)   Established the principle of using oil formulations of fungi against locust and involved a series of screenings in laboratory bioassays for virulent strains.

Phase 2 (1992–1994)   Established that the oil formulation of the virulent isolate IMI 330189 was effective in small scale field experiments. Production work at IITA to scale up the supply of conidia for trials began in earnest.

Phase 3 (1995–1998)   Larger scale field trials, toxicological, ecotoxicological and economic studies were carried out. These studies indicated that commercialisation offered the most favourable route to implementation. LUBILOSA licensed the technology to two commercial partners, (BCP Ltd in South Africa and NPP in France). Registration of the product was granted in South Africa.

The research results of LUBILOSA are summarised in Bateman (1997), Lomer *et al.* (2001) and Neethling and Dent (1998). A 3 year product stewardship phase of LUBILOSA began in 1999 to establish the product in the locust and grasshopper control markets in Africa; the programme is currently providing technical support to the commercial partners, advocating and promoting the product to key decision makers and users, and refining use strategies.

The Convention on Biological Diversity and the LUBILOSA Programme

The LUBILOSA Programme pre-dates the Convention on Biological Diversity (Rio de Janeiro 1992), however the principles and management of the Programme and the product Green Muscle® have been established in accordance with the following benefit sharing and related provisions of the Convention on Biological Diversity.

- Full participation in research (Article 15.6);
- Equitable sharing of research results and benefits from commercial and other use (Article 15.7);
- Access to and transfer of technology (Article 16);
- Exchange of information (Article 17);
- Technical and scientific co-operation (Article 18);
- Participation in research and access to results and benefits of biotechnology (Articles 19.1 and 19.2);
- Financial resources (Article 20).

Origin, distribution and access to the LUBILOSA isolate IMI 330189

The strain *Metarhizium* (*flavoviride*) *anisopliae* var. *acridum* IMI 330189 was isolated from *Ornithacris cavroisi* (Finot) (Orthoptera:Acrididae) near Niamey, Niger in 1988 by collaborators of a network from the Département de Formation en Protection des Vegetaux (DFPV), AGRHYMET Centre in Niamey (which services the nine CILSS countries). The collaborative programme, funded by DGIS, involved the exploration for pathogens in West and North Africa and used AGRHYMET as a base for studies of mortality factors affecting the Senegalese grasshoppers. Taxonomic studies have indicated that IMI 330189 is a pan-tropical isolate with characteristics identical to other isolates found as far afield as Australia and the Galapagos. Locusts are distributed throughout the Sahel ecological zone which extends unbroken across northern Africa and encompasses parts of Senegal, Mauritania, Mali, Burkina Faso, Niger, Chad, Sudan and Ethiopia. There is every reason to believe that the IMI 330189 group of isolates is also distributed throughout this same area.

The isolate IMI 330189 of *Metarhizium anisopliae* var. *acridum* is the active ingredient of the mycoinsecticide. This isolate is maintained in culture by CABI *Bioscience* and is available on request for R&D and non-commercial purposes. Recipients of the isolate are required to complete a material transfer agreement (MTA) in which the basic rights and responsibilities related to the specific materials transferred are defined; they agree to restrict its use to these purposes only, to control its further distribution, to not claim ownership of the isolate and to inform CABI *Bioscience* of any research results obtained. Such agreements enable maximum utilisation of a genetic resource for research while allowing the source to retain some control of the material. They also ensure that the provisions of the CBD are respected. The isolate can be requested for commercial purposes on the basis that at the point of commercialisation an appropriate agreement addressing sharing of benefits is negotiated.

A patent has not been sought nor is there any intention to do so for the purposes of securing rights to the sole use of the isolate IMI 330189 as the active ingredient of a mycoinsecticide for locust and grasshopper control — any such action is not possible under the patent regulations of the UK and most other countries, with the exception of the USA.

Product Development

The process of taking a fungal isolate and turning it into a marketable product involves the following steps:

- identification and characterisation of collected isolates;
- laboratory bioassay to determine virulence;
- formulation;
- storage tests;
- mass production;
- small scale application trials;
- large scale trials;
- operational level trials;
- assessments of ecotoxicology and mammalian and fish toxicology;
- scale-up of production;
- preparation of registration and submission of dossier;
- identifying commercial companies to manufacture, distribute and sell the product.

This process is complex, demanding and expensive. The exploitation of biodiversity via a commercial route is therefore not a straight-forward process that can guarantee success. The main issues confronting those wishing to take a commercial route to exploitation of biodiversity include:

- confidentiality;
- commercial, regulatory, technical and quality standards;
- identifying appropriate industrial partners;
- gathering market information;
- intellectual property issues;
- licensing of the technology to the private sector;

- disbursement of benefits accruing from the successful commercial exploitation of the product.

It is only through knowledge, experience and understanding of these issues and how they can be best addressed that we can be assured of the successful implementation of the CBD. LUBILOSA provides an example of the types of problems that need to be addressed and some of the options that are available to deal with the commercial exploitation of biodiversity.

Green Muscle®: the Product

Taxonomy

The taxonomy of the Deuteromycete genus *Metarhizium* has recently been revised by Driver *et al.* (2000) and identifies the acridid (locusts) active isolates as *Metarhizium anisopliae* var. *acridum*.

Active material

The active material of the mycoinsecticide Green Muscle® is based on conidia of the fungus *Metarhizium anisopliae* var. *acridum*. The standard isolate is IMI 330189, origin Niger. This isolate has been found to be effective against a wide range of acridids — including *Schistocerca gregaria*, *Oedaleus senegalensis* and *Locustana pardalina*.

Formulations

Laboratory assays have shown that the formulation of *Metarhizium anisopliae* conidia in oil improved the efficacy and speed of kill in comparison with water-based suspensions, especially at low humidity. The programme has developed a unique oil miscible flowable formulation (OF formulation).

Storage characteristics

Long-term storage of *Metarhizium* conidia is possible provided that the moisture content is kept low (below 6%). No appreciable loss of virulence is observed after 12 months at 30°C.

Application

LUBILOSA mycoinsecticide is compatible with all ultra-low volume (ULV) spraying equipment likely to be used for operational application. The current standard volume application rate is one litre per hectare. Lower rates of application of 0.5 l ha^{-1} have been successfully used and active ingredient application rates have been dropped from 100 g to 50 g ha^{-1} with no loss of efficacy.

Status of registration

Green Muscle® is registered in South Africa for control of Brown locust and has been recommended by the FAO Desert Locust Pesticides Panel for use in conservation and environmentally sensitive areas. Permission has been granted by CILSS for use in Sahelian countries pending a decision on registration within the biopesticide regulatory framework being developed.

Green Muscle®: the Route to Dissemination and Technology Transfer

The LUBILOSA Programme adopted a "two technologies" approach to the development and dissemination of the mycoinsecticide: a low technology, labour intensive, artisanal approach and a capital intensive, high technology, commercial approach to mycoinsecticide production.

The artisanal approach to production of the mycoinsecticide, while adequately serving the needs of the LUBILOSA research programme, proved inappropriate for scale-up and mass production. This was simply because it requires a relatively high capital input (US$100,000 per plant) in order to maintain a sufficiently high quality product, has a maximum output in practice of 2,000 ha year^{-1} (when the smallest market for the mycoinsecticide in 50,000 ha) and the spore cost of US$20 ha^{-1} compares less favourably with competitor products of US$7–12 ha^{-1} (Cherry et al., 1999). Delivery costs would elevate the cost further.

In contrast, the high technology, capital intensive commercial production has a minimum entry point into the market at 50,000 ha (for it to be an economic proposition), has a modular capability to expand production to 200,000 ha per plant (demands greater than this will be met by investment in new in-country production plants) and are expected to be able to produce spores at a more competitive price (Swanson, 1995).

Two commercial companies (Natural Plant Protection (NPP), based in France who will service West African grasshopper markets through a plant in Senegal; Biological Control Products (BCP), based in South Africa who will service southern African locust markets) have been licenced to manufacture, market and sell Green Muscle®.

The basic principle on which LUBILOSA has approached the development of Green Muscle® for locust and grasshopper control has been to meet the "public need" with a basic, workable product and system for production which has been made readily available to the public domain. However, the LUBILOSA manufacturing process was assessed to be uneconomic. LUBILOSA is transferring its technical knowledge of production and formulation to two companies who bring to bear their experience in large scale manufacturing expertise of biopesticides.

Disclosure, Sharing and Exchange of Information

In general, public disclosure of information is considered unhelpful to commercialisation of a product but the decision of whether or not to disclose information generated by a research team also has implications for individual scientists and the donors funding the research.

Public sector scientists in general

Scientists progress their careers by publishing papers in scientific journals (Dent, 1995). Hence, scientists required to keep the results of their research confidential because of the commercial value of such information can suffer from lack of promotion and acknowledgement of their capability among the wider scientific community. For this reason most scientists working in the public sector prefer to publish the results of their research if at all possible.

Research providers

Public and private research providers may have different attitudes to confidentiality of generated know-how. Traditionally for instance, the CGIAR institutes have advocated full public disclosure of all scientific activities and outputs which makes working with the private sector problematical. More recently, a special case can be made for retaining some information confidential on a "case-by-case" basis. Such retained information is viewed as "bargaining chips" to enable favourable outcomes in negotiations with private companies (Report of the CGIAR expert panel on proprietary Science and Technology; document SDR/TAC:IAR/98/7.1, CGIAR mid-term meeting 1998, Brasilia, Brazil). In contrast, Universities have increasingly, since the 1980s, become aware of the commercial potential of research and have established Business

Units to address issues of confidentiality and IPR. The terms, conditions and costs of research projects will vary according to who holds the IPR and whether or not the research provider can or cannot publish their research findings.

Donors supporting R&D projects

Most research projects funded by donor agencies are carried out via short term contracts to the research provider. The contracts usually specify the terms and conditions with regard to public disclosure of information. The donors tend to have two concerns: (i) that information generated by the research provider for public release is scrutinised to ensure it does not in any way prejudice the Government of the funding agency; and (ii) that no information generated by the research provider that is for the public good and conforms with the development aid mission of the funding agency is unreasonably withheld (see Box A).

Commercial companies

Commercial companies undertake research, development, design and implementation of processes and manufacture of products that provide them with the means by which they exist as a business. The use of confidentiality is a crucial instrument for maintaining their ability to function in competitive business situations, where it is used to prevent other companies from benefiting from know-how generated or obtained by the company. Companies protect their interests through patents, trademarks, registered designs, industrial secrets and through use of confidentiality agreements. The use of confidentiality agreements is the most common form of control exercised by companies to enable collaboration between different parties while protecting the commercial interests of those parties. Hence, commercial companies interested in exploiting particular know-how have a preference for that which has been kept confidential (as an "industrial secret" or protected by patents, etc.) and has not been disclosed to the public. The commercial value of the know-how rests in the fact that no-one else has access to the information which then provides the company with a competitive advantage that they can exploit. To this end, signing a confidentiality agreement that seeks to protect the interests of both parties (Annex A) precedes any discussion with commercial companies.

Dealing with confidentiality at project level

The interests with regard to confidentiality within a product-orientated project, funded by donors, implemented by scientists and engaging commercial companies, are at total variance. The LUBILOSA Programme has managed to marry these different interests over the course of its 10 years.

Box A: Examples of Confidentiality Clauses in Donor Contracts

CIDA/CABI Agreement no. 25471
RC/Project: 806/18880
Signed 19 November 1996

12.0 Disclosure

12.0.1 The Organization shall not disclose any matter, information or documents which may come to its knowledge or possession by reason of its participation under this Contribution unless it has received the prior written consent of CIDA.

12.0.2 The Organization shall ensure that its personnel, outside consultants or sub-contractors are bound by the provision of article 12.0.1.

12.0.3 The Organization shall refrain from any action which might be prejudicial to the friendly relations between Canada and all the countries where activities of the project shall be conducted.

DGIS/CABI Proj. no. RF017207
Signed 12 July 1996

Article VI (rights to reports/industrial property rights)

1.   All reports, maps, diagrams, designs, drawings, models, statistics and computer software produced by the Organisation under the terms of this agreement shall be properly arranged and indexed by the Organisation and kept with care for five years after the termination of this Agreement.

   Such data shall be freely available at all times to the Minister and the public at large.

The LUBILOSA Programme has maintained a policy of public disclosure of the information, results and outputs generated throughout the course of its research and development. LUBILOSA has pursued two routes to ensure that information generated is kept in the public domain. The first was to publish in scientific journals the results of the research undertaken by the programme. A total of 120 scientific papers, conference proceedings and reports have been produced by LUBILOSA scientists and collaborators that detail the techniques, approaches and results of the research, including production, formulation, storage, ecotoxicology and application. These publications which are widely available provide sufficient information to enable anyone to develop, produce on an artisanal scale, formulate and apply a basic but highly effective mycoinsecticide for locust and grasshopper control.

In addition, all collaborators have been able to contribute to and receive copies of a regular programme newsletter disseminating information about the Programme and its achievements. Representatives of national programmes have been invited to and attended numerous scientific meetings and Programme Management Meetings over the 10 year duration of the Programme.

Since publication of results in the scientific literature can take time (2 years for some journals), there was some concern among donors and LUBILOSA partners that commercial companies may patent the technology and thereby restrict its use in developing countries (CABI/LUBILOSA Programme Management Committee Minutes 1989–1992). For this reason it was decided that CABI would patent the technology that had led to the establishment of the LUBILOSA Programme (Box B). In doing so, it prevented others filing patents and this has allowed the more widespread development and use of oil formulations with Deuteromycete fungi.

The only minor exception to the general policy of complete freedom of information occurred in relation to the technical details of the more sophisticated oil miscible (OF) formulation and a limited amount of information relating to spore storage models. LUBILOSA has made public an estimated 99.5% of the information generated through its research. In practical terms, this means the "public" have access to all the information they require to allow them to develop a small scale production capability, formulate, store and apply the mycoinsecticide using the simple but robust SU formulation. This is the formulation that has a proven field performance and is recommended for use by the FAO Pesticides Referee Group. The remaining small amount of work that is not in the public domain relates to a highly sophisticated formulation, that has not been extensively tested, is not currently approved by FAO and in addition requires expensive, specialised equipment to manufacture. The OF formulation could not be produced by an artisanal approach because its manufacture requires the use of costly specialist machinery. Maintaining confidence about the technical specifications of the OF formulation does not preclude exploitation of the SU formulation of the mycoinsecticide by non-commercial producers and artisanal producers in developing countries. CABI *Bioscience* generated the confidential know-how in only Phase 3 of the programme and CABI *Bioscience* has maintained this as industrial secrets on behalf of LUBILOSA with the agreement of its partners.

Box B: Oil Formulation Technology Patent

The invention by the CABI *Bioscience* scientists, Drs C. Prior, R. Bateman and D. Moore, that Deuteromycete fungi formulated in oil permits the utilisation of fungal based insecticides in arid environments is the subject of a patent entitled "Entomopathogenic sprays" No. GB2255018B granted to CAB *International* on 15 March 1995. This technology has been exploited in the LUBILOSA Programme.

Ownership of the intellectual property

The donors contract an "implementing organisation" to undertake a programme of work. The terms and conditions with regard to the ownership and exploitation of the intellectual property generated in the execution of a programme are specified in the contracts between the donors and the implementing organisation. The main concern of donors is that the access to intellectual property generated from the R&D that they fund is not restricted in any way. This is to protect the interests of those that the donors seek to assist in developing countries.

In the normal course of events if an organisation generates know-how which has some commercial value it will licence this know-how to a commercial company to exploit. However, for the organisation to be able to do this it must be the owner of the intellectual property. Where the organisation is contracted to undertake the R&D on behalf of a donor then such a licence would need to be issued by the donor or the donor would need to transfer the rights to exploit the technology to the R&D provider. Since it would normally be considered inappropriate for a donor agency to licence directly and to manage the licensing of a commercial company to manufacture a product, this only leaves the option for transfer of the intellectual property from the donor to the R&D provider as a practical solution. The difficulties then arise that the donor has to ensure that by taking such a route they in no way compromise their mission or disadvantage those they most want to assist. Thus, any IPR transfer agreement must incorporate statements that protect the interests of the donors without unreasonably hindering the commercial exploitation and the success of the project. Where there is more than one donor involved in the funding of the project then the issues of transfer of intellectual property and its commercial exploitation are further complicated.

CABI *Bioscience* was contracted by CIDA, SDC, DGIS and DFID to implement the LUBILOSA Programme and as such is responsible to the donors for undertaking all the technical, financial, administrative and legal matters relating to the Programme. As the implementing organisation CABI signs the contracts on behalf of the LUBILOSA partners and is legally bound by the contract to only exploit the intellectual property generated in accordance with the terms and conditions set out therein (Box C).

The implications of the CABI contracts with the different donors was that CABI did not wholly control the intellectual property of the LUBILOSA programme and, hence, could not assign the rights of licence to the commercial companies to allow them to manufacture the mycoinsecticide. In addition, the donors were not obliged to maintain the know-how generated by the programme confidential (see above). To enable CABI to licence the know-how to the commercial companies the donors needed to transfer their rights to the intellectual property to CABI on terms and conditions that would guarantee their interests were maintained. To this end the donors agreed to a multi-sponsor IPR agreement which served this function. The clauses dealing with the undertakings of the sponsors included statements with regard to transfer of the intellectual property, the purposes of the transfer, confidentiality and a statement protecting the donors from claims concerning warranty of the know-how (Box D).

In return, and in order to protect the interests of the donors and those they seek to benefit the following issues were addressed in the agreement: good commercial practice, product price, product availability, donor statutory purposes and legal obligations (Box E).

Box C: Examples of Intellectual Property Rights Clauses in Donor Contracts

DGIS Proj. no. RF017207
Signed 12 July 1996

Article VI (rights to reports/industrial property rights)

5.   In the case of any inventions or other achievements accomplished in the course of activities and which could lead to the establishment of industrial property rights, the Organization shall grant, regardless of whether a patent has been granted or some other form of legal protection accorded to the invention or the achievements, free of charge, a licence or the right of application or use to organizations based in developing countries, or to an organization which performs development co-operation activities.

ODA Contract no. CNTR 96 1498A
Signed 30 December 1996
Disclosure of Information, Intellectual Property Rights and Official Secrets Act

.2 Where the Consultants are contracted to supply Services directly to ODA all reports should be addressed directly to the Project Officer. All intellectual property rights in such reports and any other documentation prepared for the purpose of performing the Services shall be the property of the Consultants.

.3 The Consultants hereby grant to ODA or the Overseas Recipient a world-wide non-exclusive irrevocable royalty free licence to use the reports and any intellectual property rights therein as described in sub-clauses 19.2.2.

Box D: Undertakings by the Sponsors

The Sponsors hereby undertake to waive their intellectual property rights and claim (described in Annex B) to the KNOW-HOW, the MATERIALS and the PRODUCTS and to cede these rights to CABI for the purposes of exploiting and licensing of the KNOW-HOW, MATERIALS and PRODUCTS to commercial companies on behalf of the LUBILOSA Programme. The Sponsors give no warranty or representations (either express or implied) relating to the intellectual property rights or KNOW-HOW, MATERIALS or PRODUCTS.

The Sponsors hereby undertake to keep confidential all commercially sensitive information provided to it by the other Party and to treat such information with the same care as it would treat its own commercially sensitive information.

Such information shall only be disclosed to such employees and agents of the sponsors as have need to know it for the purpose of carrying out the requirements of this agreement, and such employees and agents shall be made aware of and bound by this burden of confidentiality. This burden of confidentiality shall remain in force until the second anniversary of termination of this agreement, or the tenth anniversary of its effective date, whichever is the later.

Box E: Undertakings by CABI

CABI shall only exploit the KNOW-HOW, MATERIALS and PRODUCTS and TRADEMARKS ceded by the Sponsors in accordance with CABI's purpose and function (Pursuant to Article XVII, Paragraph 3; the Agreement on CABI entered into force on 4th September 1987), the terms and conditions specified in the contractual agreements made between each of the Sponsors and CABI and good commercial practice with recourse to the appropriate transfer of public sector technology to the commercial sector.
CABI undertakes to ensure through the agreements of licence with commercial companies, for the purposes of exploiting the LUBILOSA technology that the following conditions apply:

a) a reasonable price is charged for the sale of the product; the price must be competitive with other similar products to ensure general use and accessibility;
b) there is a reasonable availability of the mycoinsecticide within sponsor core countries (Second Schedule to this agreement) requiring such products;
c) the agreements of licence reflect and be in the interests of Sponsors statutory statements of purpose;
d) the commercial companies will through the mechanism of a licence agreement acknowledge that the Sponsors of the KNOW-HOW, MATERIALS and PRODUCTS shall in no event be restricted from complying with their respective statutory duties and all international legal obligations to co-operate with other international development agencies and organisations in relation to the KNOW-HOW, MATERIALS and PRODUCTS.

Commercial Company Collaboration

The interaction with a commercial company is a two way process. The commercial company needs to be convinced that the product is commercially viable and the licencee needs to be sure that the commercial company has the where-with-all to register, manufacture, market and sell the product to the required standards and price.

Key issues for both parties to collaborate on mycoinsecticide commercialisation include:

1. Product specification: sufficiently broad spectrum for there to be a sizeable market for the product, high virulence, good speed of kill, good storage capability, use of a conventional formulation utilising existing application equipment;

2. Production: utilisation of an established production processes, conventional packaging and storage;

3. Markets and demand: a number of large, regular, well-established markets, few competitive products, a specific product advantage for which there is an established demand or a well-defined niche market presently unexploited that provides an economically attractive opportunity;

4. Distribution and sales: product must fit within existing networks of distribution; wholesale and retail sales outlets and mechanisms need to be well-established;

5. Toxicology and ecotoxicology: product should ideally be environmentally friendly and have low vertebrate toxicity;

6. Registration: product should require first tier testing only and enter a fast track registration process;

7. Economics: favourable toxicological attributes reduce development and registration costs, use of an established production process reduces developmental costs. Production, packaging, storage and transport costs need to be low, price needs to be competitive.

Green Muscle®: positive attributes

Green Muscle® has a large number of the positive characteristics outlined in point 1 which are likely to lead to its successful commercialisation as a biopesticide. Green Muscle® has:

- high virulence and prolonged field efficacy;
- relatively good storage capability;
- uses conventional application equipment;
- involves a simple, low-cost production process;
- standard packaging and storage requirements for a biological product;
- safety to operators;
- minimal environmental effects.

Across the whole spectrum of locusts and grasshoppers for which control is considered necessary in Africa, there is a well-established, regular market with additional markets available in locust outbreak years. Demand for an environmentally friendly, low-hazard product is high. The distribution and sales networks are the same for both chemicals and biopesticides.

Registration costs are expected to be low and fast track. Given current estimates available, the mycoinsecticide can be priced competitively with chemical pesticides. Modifications to the mycoinsecticide formulation have reduced dosage rates and hence costs of application, making the mycoinsecticide even more competitively priced. The persistence of the product makes repeated application unnecessary which also reduces costs (Langewald *et al.*, 1999).

Green Muscle®: negative attributes

The characteristics on which Green Muscle® might be considered at odds with successful commercialisation include specificity, relatively slow speed of kill and a perceived irregular market size and demand.

- **Specificity**: is largely a trade-off between market size and potential for environmental damage. Broad spectrum pesticides often have a detrimental impact on beneficial insects and other arthropods but enable targeting of a broad range of pests which increases the market scope. Highly specific insecticides by definition have little detrimental impact on beneficial insects but a restricted range of target pests that reduces market scope and economic viability. Green Muscle® based on the strain IMI 330189 is specific to locusts and grasshoppers on the whole, but may have an application against a limited range of other pest species.
- **Slow speed of kill:** chemicals used for locust and grasshopper control in recent years have a fast knockdown and fast speed of kill. By comparison, the mycoinsecticide takes on average 6 days before any mortality is observed and

10–15 days to reach maximum mortality due to direct hit. This relatively slow speed of kill represents a barrier to commercialisation in the sense that there is a need to overcome the perception that associates slow speed with ineffectiveness. The relatively slow speed of kill of some stomach poisons such as substituted ureas or IGRs has not prevented their widespread uptake and use, hence slow speed of kill is not on its own considered a major threat to the commercialisation of Green Muscle®.

- **Storage requirements**: a pesticide must be capable of long term storage under a range of conditions to survive the process of distribution, sale and use. The periods of exposure to fluctuating storage conditions will be determined by demand, speed of delivery and time to ultimate use. The "rule of thumb" for chemical insecticide storage capability is 18 months at 30°C and for biopesticides 6 months at 20°C. Green Muscle® storage capability is at present > 12 months at 30°C which is close to chemical standards and significantly better than most other biopesticide products. The final commercial product, incorporating all of the most recently generated research results, may well achieve storage properties as good as those of chemical insecticides.

- **Market size and demand**: the potential commercial success of a product is dependent on exploiting an identifiable market with a sustainable demand, sufficient to recover all costs and achieve an acceptable level of profit. Large markets and a regular demand represent attractive opportunities for commercial exploitation. The market for pesticides for the control of desert locust (*Schistocerca gregaria*) is large but highly irregular, although demand is high for an environmentally friendly product. Markets for other locust species such as *Dociostaurus moroccanus* in Morocco, *Locusta migratoria* in Madagascar, *Anacridium* spp. (Tree locusts) in Sudan and *Locusta pardalina* (Brown locust) in southern Africa are smaller but with a more regular demand. A large pesticide market with a regular demand exists for grasshopper control in West Africa.

The *Schistocerca* market size is large enough but not regular enough to be commercially viable on its own. The West African grasshopper market is commercially viable but dependent on Japanese KR2 funding which will probably be available for the next 5–10 years. The Moroccan, Tree and Brown locust markets are small but commercially sustainable over the next 5–10 years. Hence, LUBILOSA targeted the Brown locust market, followed by the West African grasshopper market as a commercially sustainable means of accessing the market for *Schistocerca* control when outbreaks occur and demand is high.

Approaches to commercial companies

In approaching commercial companies for the purpose of entering into partnerships, LUBILOSA has taken account of their size since this has many implications for the type and value of the arrangement that can reasonably be expected. For the sake of brevity companies are considered here as extremes, large multinational and small to medium enterprises (SMEs; EU classification < 50 employees).

Advantages of large multinationals	Disadvantages of large multinationals
• presence, i.e. selling power in the market • market penetration and expertise • financial resources (for registration, publicity, marketing) • logistic resources (established networks, distributors, retailers)	• large inertia • time taken to establish relevant contacts, gain approval for partnerships • tendency to "sit on" niche market products. • unlikely to obtain a non-exclusivity agreement • will not consider markets of less than £15 million
Advantages of SMEs	**Disadvantages of SMEs**
• good knowledge of specific niche markets (< £50 million) • flexibility and speed of action • a hunger for new products and new opportunities • will accept non-exclusivity agreements • will exploit new products quickly if resources available	• lack of resources and capital investment capability • presence in market place limited • lack of influence, limited networks, etc.

Time is an important constraint in project R&D. Given the problems that could arise with arranging collaboration in the time scale and on terms agreeable to LUBILOSA, the development of non-exclusive agreements with a number of SMEs was considered preferable to attempting to obtain agreements with large multinationals. In addition, most of the locust and grasshopper markets are relatively small and unlikely to interest a large multinational unless other markets were available for the mycoinsecticide, which, at present, they are not. Although the *Schistocerca* (Desert locust) market fluctuates enormously as a whole, it is likely that the mycopesticide demand would be more stable. Green Muscle® would be used principally in a preventative context; major flying swarms and plagues would still be treated with chemicals. Hence, it is anticipated that there would be a stable consumption of 40–60% of the market between outbreaks, falling to perhaps 10–20% (but increasing in absolute amounts) in plague years.

Commercial company interest in biopesticides

During the 1960s and 1970s, large fermentation companies (Abbott, Sandoz, IMC) were the industry leaders. Their production costs for these biopesticides were low and the key challenge was to make high quality products and to market them well. Unfortunately these companies were not very successful so that now only Abbott retains a significant share of the market. In the 1980s, the picture began to change. Small biotech companies (Mycogen, Ecogen, BioSys, Crop Genetics, etc.) began to appear, fuelled by venture capital funding and promising greatly expanded markets for biopesticides. Although investment was forthcoming the technical problems concerning some of the products were never properly resolved. This led to a process of mergers and take-overs — so today few of these companies exclusively market biopesticides, making their income from a range of other associated products. In the 1980s a number of the agrichemical companies (Monsanto, Ciba Geigy, DuPont, Amercan Cyanimid) invested in biopesticide research based on the assumption that a small investment in R&D would yield substantial benefits. The markets however, were smaller than predicted, public demand for chemical free products was overestimated and biotechnology could not radically improve biopesticide performance. In addition, these companies and others investing in biopesticides attempted to apply the centralised production, global distribution model of the agrichemical products to biopesticides (Dent and Waage, 1999). The model is inappropriate for biopesticides given their specificity, niche market and reduced storage capability compared to chemicals. Given this, in the late 1990s a small number of highly specialised biopesticide companies with niche market products distributed nationally or regionally have shown some success. It is with such companies that the future success of biopesticides now lies.

Which companies?

There are a number of companies that already have mycoinsecticide products on the market and a number of others have the capability to produce and market them. The decision as to which companies LUBILOSA should approach depended on their ability to meet the following criteria:

1. an SME;
2. production capability that can be readily adapted to *Metarhizium*;
3. "access to donor funding", particularly KR2 Japanese aid;
4. an existing distribution system in appropriate regions (primarily West and East Africa; Southern Africa);
5. access to capital;
6. track record in registering, producing, marketing and selling biopesticides;
7. willingness to enter into an appropriate licensing agreement.

In addition the LUBILOSA Expert Advisory Committee (EAC: independent experts advising LUBILOSA scientists) recommended that the Programme broaden the scope of potential interested parties by producing and distributing a leaflet outlining the opportunity offered for investment by Green Muscle®. One hundred leaflets were produced and distributed — responses to these were followed up but no new leads were obtained.

Companies with access to the acridid market

West Africa

The market for acridicides in Africa accounts for less than 5% of the total pesticide market (US$1 billion) and is dominated by Japanese KR2 acridicide aid since 1989 including that supplied through FAO. Such a predominance in the market could create distortions in supply for chemical and mycoinsecticides. Swanson (1995) concludes that if the Japanese are not included in the development or exploitation of Green Muscle® then the prospects for its success are decreased.

As a leading donor of pesticides for acridid control in Africa, Japan could purchase Green Muscle® for acridid control through its KR2 aid programme. In order to do this a Japanese owned trading company would need to places a bid for KR2 on the basis of a request from a local representative in the trader company or neighbouring country (Abidjan is acceptable for most of West Africa outside of Senegal). In addition, the product must be produced and reformulated in Japan or an OECD country, must meet WHO and FAO safety and environmental guidelines and be registered for use in Japan. Hence, to have any chance of qualifying for KR2, Green Muscle® needs to be linked through collaboration to a Japanese owned or partially owned company. At present no Japanese companies are involved in the development and production of mycoinsecticides but they own subsidiary companies that are potentially interested in this market, namely Natural Plant Protection (a subsidiary of Calliope, which is a subsidiary of Nichimen).

Calliope is owned (100 %) by Nichimen and has a 51% holding in NPP. The other shareholders in NPP are Sumitomo Corp. (19.5%), Agrolinz Melamin (3.75%), M. Guillon (Managing Director: 16.25%) and other private investors (9.50%). NPP is a French Company with capital of 16 million FF that has developed and is marketing a whole range of microbial products (viruses, bacteria and fungi) including Ostrinil® a *Beauveria bassiana* based mycoinsecticide for the control of European corn borer and Betel® for the control of sugar cane white grub. The company possesses a modern fungus production plant in France and ReUnion and is considering building another in Senegal, West Africa. NPP has access to the acridid market through its parent company Calliope which has gross annual sales of pesticides in Africa of US$100 million, of which one quarter is for locust and grasshopper control. Calliope has a 20% share in SPIA, a Senegalese reformulation company. Participation in tenders for KR2 funding would thus, in principle, be possible through NPP and Calliope.

South Africa

Biological Control Products Pty Ltd is an SME based in Durban, South Africa whose core business is to manufacture, market and sell biological control agents for use in the control of plant pathogens. Their main product is the nematicide based on the fungus *Paecilomyces lilacinus* for the control of nematodes of tomatoes, potatoes and groundnuts which is registered in South Africa as PL Plus. The *Paecilomyces* production plant in Pinetown utilises a similar solid substrate system to that required for industrial scale production of *Metarhizium*. The production facility was commissioned in December 1997. The capital investment for BCP has been financed through equity contributions of investors and a grant from the Industrial Corporation, South Africa. If the market for *Metarhizium* extends beyond that for control of Brown locust then the company will expand their manufacturing plant.

A regional presence

The nature of the funding for LUBILOSA focuses attention on the need to promote the development of production of the mycoinsecticide within Africa. This will boost local involvement, employment and the economy in those regions where production plants are established. There are few commercial companies that have production capability within Africa relevant to mycoinsecticide manufacture. Calliope/NPP may develop a production plant in Senegal; BCP have a production facility in Pinetown.

The basis of LUBILOSA collaboration with commercial companies

Prior to entering into any discussions with commercial companies, confidentiality agreements were signed to protect LUBILOSA IPR. Licensing agreements were negotiated on the basis of the following:

1. a non-exclusive basis incorporating a specific geographical, pest species or cropping system jurisdiction;
2. transfer of liability;
3. an advance payment ("licence fee") (scale dependent on company and royalty);
4. a royalty payment in the range 2.5–7.5%;
5. a non-assignable agreement and acceptable accounting procedures.

In return, LUBILOSA provides the commercial company with:

1. toxicological and ecotoxicological information relevant to, but not necessarily wholly inclusive of, requirements for registration of the product;
2. relevant efficacy data and results;
3. know-how and expertise to assist in the production process, registration, labelling and marketing of the product.

The IPR transfer agreement and the licensing agreement with the two commercial companies included clauses to safeguard the interests of the donors. In particular with warranties with regard to acknowledgement of the donors and their statuatory obligations and product liability and use (Box F). The licensing agreement between CABI and the two commercial companies includes clauses that reflect the requirements of these warranty and liability statements (Box G).

Sharing of benefits arising from LUBILOSA

The participatory approach to isolate collection and biopesticide development which has characterised the LUBILOSA Programme clearly reflects the benefit sharing and related provisions articulated in the CBD.

A large number of agencies and organisations have participated in and contributed to the development of the LUBILOSA mycoinsecticide. It is essential that every effort is made to ensure the adoption of this technology by relevant groups in Africa. At the same time an appropriate mechanism is required to distribute the benefits arising from the technology.

Benefits arising from the development of the mycoinsecticide include:

- access to the technology and the environmental, economic and social benefits that accrue from its use;
- capacity building through the LUBILOSA training programme;
- royalties generated from sale of the mycoinsecticide.

As discussed in the CBD and in several interpretative articles (Iwu, 1996) the principal issue is not the disbursement of the royalty payments (which in the case of LUBILOSA are likely to be small) but the participation in, and access to, the technology and research developed. Access to the technology has been dealt with above and includes the elements of unhindered access to the public domain information and to the use of the technology through our ability to make the mycoinsecticide available Africa-wide via the commercial route.

LUBILOSA has trained 40 scientists and technicians in insect pathology and the spore mass production process, ten scientists have been awarded PhDs (five of the scientists are from African countries) and the national programmes have participated in and developed the skills necessary to carry out experimental field

Box F: Warranty and Liability Clauses in the Multi-sponsor IPR Transfer Agreement

4. Warranties

The licensing agreements between CABI and the commercial companies shall acknowledge the Sponsors of the KNOW-HOW, MATERIALS and PRODUCTS and shall in no event restrict the Sponsors from complying with their respective statutory duties and all international legal obligations.

5. Liabilities

Inasmuch as the safety of the PRODUCTS is entirely dependent on the method of their production and subsequent use, the Sponsors accept no liability whatsoever in respect of the production and use either of the MATERIALS (whether originating from itself or from CABI or the commercial companies) or of the PRODUCTS.

Box G: Warranty and Liability Clauses in the Licensing Agreements between CABI and the Two Commercial Companies

7. Warranties

7.1 CABI hereby warrants that it is the legitimate owner and proprietor of the KNOWHOW and MATERIALS and that disclosing them and licensing them to BCP will not to the best of its knowledge infringe the rights of any third party.

7.2 BCP hereby warrants that it is not the owner and proprietor of, or the licencee of, any other knowhow, materials or products which would conflict with or in any way prevent its ability to operate this licence in good faith.

7.3 BCP acknowledges that the funders of the KNOWHOW (Canadian International Development Agency (CIDA), Swiss Development Cooperation (SDC), Directorate General for International Cooperation (DGIS) (Netherlands), and Department for International Development (DFID) (UK)) shall in no event be restricted from complying with their respective statutory duties and all international legal obligations.

8. Liabilities

Inasmuch as the safety of the PRODUCTS is entirely dependent on the method of their production and subsequent use, CABI accepts no liability whatever in respect of the production and use either of the MATERIALS (whether originating from itself or from BCP) or of the PRODUCTS, except to the extent that any serious personal injury or death resulting from such use can be demonstrated to be the direct result of negligence on the part of CABI.

trials using a mycoinsecticide. Hence, national programmes are benefiting through an increased and independent capacity to develop mycoinsecticide technology.

The reason for licensing the production, marketing and sale of the mycoinsecticide to two commercial companies was to ensure that the product Green Muscle® could be produced on a sufficient scale at appropriate international standards of quality and performance for purchase and use, largely by international and national locust control programmes.

The licences issued to the companies include clauses that ensure the following conditions apply (see Box E):

- good commercial practice with recourse to the appropriate transfer of public sector technology to the commercial sector;
- a reasonable price is charged for the sale of the product; the price must be competitive with other similar products to ensure general use and accessibility;
- there is a reasonable availability of the mycoinsecticide within sponsor core countries requiring such products;
- the monies generated from the commercial exploitation of the mycoinsecticide shall be credited to a Trust Fund and used in accordance with the declared objectives of the Fund for disbursement within Africa.

Royalties will be generated by the sale of the mycoinsecticide but the amounts of money generated are unlikely to be large. If these royalties are split between all the agencies and organisations involved in the LUBILOSA Programme, then the amounts paid to each will be insignificant and of little practical value. For this reason, it was proposed that the moneys generated from royalties on the sale of the mycoinsecticide in Africa be accrued in a Trust Fund. The purpose of the Trust will be to support collaborative initiatives associated with promoting biopesticides development and use in Africa (Box H).

The Trust Fund document specifies the purposes and principles of disbursement, the Trustee and powers, the Trust account, duration and taxation issues (see Annex C).

Box H: Clauses Dealing with the Trust Fund in the Multi-Sponsor IPR Agreement

4. Warranties

The Sponsors agree to establish an appropriate FUND for the accumulation and disbursement of monies obtained from royalties and licence fees on the sale of the mycoinsecticide in the format set out in Schedule 3.

Schedule 3 of the IPR agreement

The LUBILOSA Trust Fund have been established in accordance with the following benefit sharing and related provisions of the Convention on Biological Diversity:

- Article 1
- Article 15.6 (*full participation in research*)
- Article 15.7 (*equitable sharing of research results, and benefits from commercial and other use*)
- Article 16 (*access to, and transfer of technology*)
- Article 17 (*exchange of information*)
- Article 18 (*technical and scientific co-operation*)
- Articles 19.1 and 19.2 (*participation in research and sharing of results and benefits from biotechnology*)
- Article 20 (*financial resources*)

All monies received from licensing Green Muscle® and any down stream products world-wide (including licence fees and royalties) shall be paid by CABI *Bioscience* into the LUBILOSA Trust Account and used for the purposes of the LUBILOSA Trust.

Conclusions

The implementation of the CBD, the process of commercially exploiting biodiversity, particularly transferring public sector funded technology to the private sector creates a number of areas of contention for international development assistance agencies. Having traditionally shied away from such public/private sector collaborations there are few established protocols and procedures available. The LUBILOSA Programme provides a case study of the steps in the process of commercial exploitation of biodiversity and of the means by which a number of critical issues were resolved. The options selected within LUBILOSA were those most appropriate to the particular circumstances of this programme and may not be wholly or generally applicable.

Similar cases are likely to arise with the increasing demand for biologically based plant protection technologies (OTA, 1995). Often, as with LUBILOSA, much public money goes into the research but the profits and motivation available to the companies commercialising the products may be minimal.

Nevertheless, this LUBILOSA case study does clearly demonstrate that where the commercial route to dissemination of a technology is the most appropriate to ensure widespread availability of that technology, then it is possible for all stakeholders involved to find solutions which make access to the benefits of commercialisation possible.

Recommendations

1. Define at the outset of each development assistance R&D project whether a product will result from project outputs. Where this is the case, consider:

a) the implications for exploitation;
b) the non-commercial and commercial routes available;
c) the market potential of the potential product;
d) needs for commercial advice and input;
e) potential links with commercial companies.

2. Ensure each project establishes a publication policy and scrutinises all R&D outputs for potentially commercially exploitable know-how.

3. Ensure that all information that is not required to commercialise the product is made freely available.

4. The collection of the LUBILOSA isolate pre-dated the CBD, but future projects will need to ensure that the collection and use of any genetic resource is undertaken with the prior informed consent of and on mutually agreed terms with the source country (or institutions within the source country). Contractual agreements on genetic resources with the source country to include issues of research topics, organisms to be studied, ownership and conditions of material transfer (see also 5 below), research investment, patent rights and responsibilities, duration of agreement, and the type, amount and recipients of benefits should be considered (Rosenthal, 1997).

5. Standardise Material Transfer Agreements (MTAs) which define the basic rights and responsibilities relating to the specific materials transferred. An appropriate model is the CGIAR germplasm exchange MTA (see Annex C) but with appropriate modification for the possibility of a subsequent commercial implementation route.

6. Establish the principle of confidentiality agreements as part of the collaborative process.

7. Engage commercial companies as early in the product development process as possible.

8. In multi-donor projects thought needs to be given in the first year as to how the IPR issues are to be dealt with. Multi-sponsor agreements provide a relatively simple solution to the problem.

9. Identify and establish clear unequivocal mechanisms for the disbursements of benefits arising from commercial exploitation which should include research collaboration, access to the final product and monetary benefits derived from its licensing and sale.

References

Bateman, R.P. (1997) The development of a mycoinsecticide for the control of locusts and grasshoppers. *Outlook on Agriculture* 26, 13–18.

Cherry, A., Jenkins, N., Heviefo, G., Bateman, R.P. and Lomer, C. (1999) A West African pilot scale production plant for aerial conidia of *Metarhizium* spp. for use as a mycoinsecticide against locusts and grasshoppers. *Biocontrol Science and Technology* 9, 35–51.

Dent, D.R. (1996) Research specialisation: a constraint to integration. In: Waibel, H. and Zadocks, J.C. (eds) *Institutional Constraints to IPM.* Pesticide Policy Project Series No. 3, pp. 21–25.

Driver, F., Milner, R.J. and Trueman, W.H. (2000) A taxonomic revision of *Metarhizium* based on a phylogenetic analysis of ribosomal DNA sequence data. *Mycological Research* 104(2), 135–151.

FAO (1989) Report of the meeting on desert locust research. "Defining future research priorities" 18–20 October 1988, FAO, Rome.

Greathead, D.J. (1992) Biological control as a potential tool for locust and grasshopper control. In: Lomer, C.J. and Prior, C. (eds) *Biological Control of Locusts and Grasshoppers.* CAB International, Wallingford, pp. 4–7.

Iwu, M.I. (1996) Implementing the Biodiversity Treaty: how to make international cooperative agreements work. *TIBTECH* 14, 78–83.

Langewald, J., Ouambama, Z., Mamadou, A., Peveling, R., Stolz, I., Bateman, R., Blanford, S., Arthurs, S., Attignon, S. and Lomer, C. (1999) Comparison of an organophosphate insecticide with a mycoinsecticide for the control of *Oedaleus senegalensis* (Orthoptera: Acrididae) and other Sahelian grasshoppers at an operational scale. *Biocontrol Science and Technology* 9, 199–214.

Lomer, C.J., Bateman, R.P., Langewald, J., Johnson, D. and Thomas, M. (2001) Biological control of locusts and grasshoppers. *Annual Review of Entomology* 46, 667–702.

Neethling, D.C. and Dent, D.R. (1998) *Metarhizium anisopliae*, isolate IMI 330189: a mycoinsecticide for locust and grasshopper control. In: *Proceedings of the 1998 Brighton Conference, Pests & Diseases, 16–19 November 1998*, Vol. 1, pp. 37-42.

OTA (US Congress, Office of Technology Assessment) (1995) *Biologically Based Technologies for Pest Control.* OTA-ENV-636. US Government Printing Office, Washington DC.

Rosenthal, J.P. (1997) Equitable sharing of biodiversity benefits: agreements on genetic resources. In: *Investing in Biological Diversity, the Cairns Conference, Proceedings of the OECD International Conference on Incentive Measures for the Conservation and Sustainable Use of Biological Diversity, Cairns, Australia, 25–28 March 1996.*

Swanson, D. (1995) *Economic Viability of Mycopesticides for Acridid Control in Africa*. LUBILOSA Report, December 1995, p. 108.

The LUBILOSA web site is at: www.lubilosa.org

Annex A: Model Confidentiality Agreement

CONFIDENTIALITY AGREEMENT

THIS AGREEMENT (hereinafter "Agreement") is made on **[date]** between **[name of first party]** whose address is **[first party's address]** of the one part and of the other part.

[Consultant's name and address]

WHEREAS both Parties to this Agreement (hereinafter "the Parties") accept the following terms and conditions on which the Parties are prepared to communicate directly or indirectly to each other certain Confidential Information (as hereinafter defined) relating to carrying out the duties and responsibilities of **[title of consultancy]** (the Field of Discussion).

1.   References in this Agreement to "the Parties" include references to any associated institution, parent, subsidiary and associated companies of the Parties.

2.   "Confidential Information" means for the purpose of this Agreement any trade secrets, knowledge, information, data or experience and any business and marketing plans and projections, arrangements and agreements with third Parties, customer information including names of suppliers customers, any ideas whether reduced to a material form or otherwise, plans and models, any processes, including but not limited to technical and management processes, any drawings computer programs, instructions, designs, inventions, discoveries and intellectual property or reports thereon and copies thereof (whether technical, economic or commercial or of any other description) in any way connected with the Field of Discussion referred to above which is disclosed hereunder (whether orally or in writing) or which is obtained by the party receiving Confidential Information (the Recipient) as a result of the Recipient's presence at any premises of the party

disclosing Confidential Information (the Discloser), whether by observation of any plant or equipment of the Discloser or otherwise, or from any communications with the Discloser.

3.   The Recipient warrants and undertakes that, other than for the purpose of carrying out the duties and responsibilities of **[title of consultancy]**, at all times it will not:

a)   disclose or permit to be disclosed to any natural or legal person any Confidential Information;

b)   use the Confidential Information for any purpose other than for the benefit of the Discloser during or after the term of this Agreement; or

c)   appropriate, copy, memorise or in any manner reproduce or reverse engineer any of the Confidential Information.

4.   Such restrictions on disclosure and use of Confidential Information shall not apply to Confidential Information which the recipient can produce documentary evidence to show:

a)   is in the public domain at the date of commencement of this Agreement; or becomes part of the public domain, otherwise than through the Recipient, its officers, employees or agents; or

b)   is already in the possession of the Recipient at the time of receipt from the Discloser or is or becomes lawfully available to the Recipient on a non-confidential basis from a third party entitled to disclose it.

5.   The Recipient shall return immediately any and all Confidential Information including all copies thereof, howsoever recorded or embodied (including any magnetic media), on the request of the Discloser.

6.   Notwithstanding the foregoing, in the event of any breach of this Agreement by the Recipient, the Recipient shall indemnify and save the Discloser harmless from and against any costs, damages, loss or liability of any kind (including legal costs and disbursements in defending or settling any claim or demand made by a third party) howsoever suffered or incurred by the Discloser.

7.   The Recipient hereby assigns to the Discloser all of its present and future right (including the right to sue for any infringement of any such intellectual property right), title and interest in respect of the Confidential Information throughout the world in and to:

a) all inventions, discoveries, and designs (whether or not registrable as patents) including any invention of or developments or improvements to the subject matter of the Confidential Information;

b) the copyright in all works; and

c) all Confidential Information.

8. The Recipient shall make full disclosure to the Discloser of all inventions, discoveries, designs and works referred to in clause 7 created, made or written during or after the term of this Agreement.

9. The Recipient acknowledges that, by virtue of the Discloser's ownership of the Confidential Information and the intellectual property rights assigned to it under this Agreement, the Recipient may not now or at any time in the future use or exploit the Confidential Information or the said intellectual property without the expressed written permission of the Discloser except in the performance of its obligations under this Agreement.

10. Nothing contained in this Agreement shall prejudice or affect or be deemed to confer on the Recipient rights under any patent or like protection at any time owned or controlled by the Discloser.

11. This Agreement shall be subject to English law.

SIGNED by   --

On behalf of **[name of organisation of first party if applicable]**

Date   --

SIGNED by   --

On behalf of **[name of organisation of second party if applicable]**

Date   --

Annex B: Multi-Sponsor Intellectual Property Rights Agreement

MEMORANDUM OF UNDERSTANDING
INTELLECTUAL PROPERTY RIGHTS AGREEMENT

THIS MEMORANDUM OF UNDERSTANDING (hereinafter MOU) is made on
. . . . between the sponsors (hereinafter "the Sponsors") of the
LUBILOSA (Lutte Biologique contre les Locustes et Sauteriaux; Biological
Control of Locusts and Grasshoppers) Programme, the:

Swiss Agency for Development Co-operation (SDC) whose address is:

> *Environment, Forests and Energy Division, Swiss Agency for
> Development Co-operation, Federal Ministry of Foreign Affairs,
> Eigerstrasse 73, CH-3003 Bern, SWITZERLAND;*

**The State of the Netherlands, represented by the Minister for Development
Co-operation** whose address is:

> *Netherlands Ministry of Foreign Affairs, Department for Rural and
> Urban Development, Bezuidenhoutseweg 67, 2500 EB The Hague,
> NETHERLANDS;*

**The Secretary of State for International Development at the Department for
International Development (DFID) (UK)** whose address is:

> *DFID, 94 Victoria Street, London SW1E 5JE, UK;*

and **CAB *International***, a non-profit international inter-governmental
organisation established by an international treaty level agreement registered with

the United Nations on 11 January 1988 ("CABI"), operating through CABI *Bioscience* based at Silwood Park, Buckhurst Road, Ascot, Berks SL5 7TA, UK.

1. Whereas:

A. All Parties to this MOU (hereinafter "the Parties") accept the following terms and conditions on which the Sponsors are prepared to cede intellectual property rights relating to the LUBILOSA Programme mycoinsecticide to CABI in return for the establishment of the LUBILOSA Trust Fund and payments thereto.

B. CABI is contracted by the Sponsors to manage the LUBILOSA Programme. The LUBILOSA Programme (Lutte Biologique contre les Locustes et Sauteriaux) is a collaborative programme established in 1989 to research and develop an experimental mycoinsecticide containing spores of the insect pathogenic fungus *Metarhizium* for use in locust and grasshopper management.

C. The implementation of the LUBILOSA Programme has generated know-how (the "KNOW-HOW") relating to the propagation and cultivation of the fungus *Metarhizium anisopliae* var. *acridum* and the production of formulations thereof for use as a mycoinsecticide for the control of locusts and grasshoppers (the fungus and the known formulations together being referred to as the "MATERIALS"), toxicological and ecotoxicological data relevant to the registration of the mycoinsecticide as a pesticide product, the KNOW-HOW and MATERIALS being more fully described in the First Schedule to this MOU.

D. CABI is currently the registered proprietor of the UK registered trademark "Green Muscle" (the "TRADEMARK") which may be used to describe the MATERIALS and any products based on or incorporating the MATERIALS (the "PRODUCTS"), and also intends to apply for registration of the same trademark in other territories.

2. Undertakings by the Sponsors

The Sponsors hereby undertake to waive their intellectual property rights and claim (if any) (described in Annex A) to the KNOW-HOW, the MATERIALS and the PRODUCTS and to cede these rights to CABI for the purposes of exploiting and licensing of the KNOW-HOW, MATERIALS and PRODUCTS to commercial companies on behalf of the LUBILOSA Programme. The Sponsors give no warranty or representations (either express or implied) relating to the intellectual property rights or KNOW-HOW, MATERIALS or PRODUCTS.

The Sponsors hereby undertake to keep confidential all commercially sensitive information provided to it by the other Party, and to treat such information with the same care as it would treat its own commercially sensitive information.

Such information shall only be disclosed to such employees and agents of the Sponsors as have need to know it for the purpose of carrying out the requirements of this agreement, and such employees and agents shall be made aware of and bound by this burden of confidentiality. This burden of confidentiality shall remain in force until the second anniversary of termination of this agreement by all Parties, or the tenth anniversary of its effective date, whichever is the later.

3.   Undertakings by CABI

CABI shall only exploit the KNOW-HOW, MATERIALS and PRODUCTS and TRADEMARKS ceded by the Sponsors in accordance with CABI's purpose and function (Pursuant to Article XVII, Paragraph 3; the Agreement on CABI entered into force on 4th September 1987), the terms and conditions specified in the contractual agreements made between each of the Sponsors and CABI (excluding those clauses specified in Annex A) and good commercial practice with recourse to the appropriate transfer of public sector technology to the commercial sector.

CABI undertakes to ensure through the agreements of licence with commercial companies, for the purposes of exploiting the LUBILOSA technology, that the following conditions apply:

a)   a reasonable price (by reference to local market conditions) is charged for the sale of the PRODUCTS; the price must be competitive with other similar products to ensure general use and accessibility;

b)   there is a reasonable availability (by reference to local market conditions) of the mycoinsecticide within sponsor core countries (Second Schedule to this agreement) requiring such PRODUCTS;

c)   the agreements of licence reflect and be in the interests of the Sponsor's statutory statements of purpose;

d)   the commercial companies will through the mechanism of a licence agreement acknowledge that the Sponsors of the KNOW-HOW, MATERIALS and PRODUCTS shall in no event be restricted from complying with their respective statutory duties and all international legal obligations to co-operate with other international development agencies and organisations in relation to the KNOW-HOW, MATERIALS and PRODUCTS;

e) the monies generated from the commercial exploitation of the mycoinsecticide shall be credited to the LUBILOSA Trust Fund (the "FUND") (Third Schedule to this agreement) and used in accordance with the declared objectives of the FUND. CABI shall negotiate the maximum level of licence fees and royalties reasonably available from the commercial companies (by reference to prevailing market conditions) for the benefit of the FUND and (where applicable) the Sponsors; and that

f) CABI will include provisions in the agreements of the licence to fully protect the intellectual property rights in the KNOW-HOW, MATERIALS and PRODUCTS.

4. Warranties

The licensing agreements between CABI and the commercial companies shall acknowledge the Sponsors of the KNOW-HOW, MATERIALS and PRODUCTS and shall in no event restrict the Sponsors from complying with their respective statutory duties and all international legal obligations.

The Sponsors agree to establish an appropriate FUND for the accumulation and disbursement of monies obtained from royalties and licence fees on the sale of the mycoinsecticide in the format set out in Schedule 3. CABI shall attend meetings of the Disbursement Committee of the FUND if requested in a advisory capacity.

5. Liabilities

Inasmuch as the safety of the PRODUCTS is entirely dependent on the method of their production and subsequent use, the Sponsors accept no liability whatsoever in respect of the production and use either of the MATERIALS (whether originating from itself or from CABI or the commercial companies) or of the PRODUCTS. CABI shall indemnify and keep indemnified the Sponsors from any claims by third parties against the Sponsors in connection with the MATERIALS or the PRODUCTS.

6. Termination

This MOU shall come into force on the date first hereabove written and shall continue in force until terminated by written agreement of all Parties.

Any of the Sponsors may, on 28 days written notice to all Parties, withdraw from the MOU whereupon this MOU will terminate with respect to the Sponsor withdrawing . In such event, the MOU will continue in full force and effect between all the remaining Parties. Termination of this MOU with respect to all Parties or any single Sponsor shall be without prejudice to the rights of any Party accrued prior to such termination.

7. Sole and Entire Memorandum of Understanding

This MOU supersedes any prior agreement or understanding whether written or oral in respect of the KNOW-HOW, MATERIALS, PRODUCTS and TRADEMARKS.

8. Applicable Law

Any issue arising in respect of this MOU which cannot be resolved amicably between the parties shall be referred to the Permanent Court of Arbitration, in the Hague.

9. Signatories

Signed	Signed
Date	Date
Name	Name
Position	Position
For and on behalf of Department	*For and on behalf of Swiss*
for International Development (UK)	*Development Co-operation*
Signed	Signed
Date	Date
Name	Name
Position	Position
For and on behalf of Dept. for	*For and on behalf of*
Rural Development, Ministry of	*CAB International*
Foreign Affairs (Netherlands)	

Annex C: CGIAR Material Transfer Agreement

Text of this agreement can be found at www.cgiar.org/whatis.htm

MATERIAL TRANSFER AGREEMENT

(MTA)

The material contained herein is being furnished by [Centre] under the following conditions:

Designated Germplasm

[Centre] is making the material described in the attached list available as part of its policy of maximising the utilization of genetic material for research. The material was either developed by [Centre]; or was acquired prior to the entry into force of the Convention on Biological Diversity; or if it was acquired after the entering into force of the Convention on Biological Diversity, it was obtained with the understanding that it could be made freely available for any agricultural research or breeding purposes.

The material is held in trust under the terms of an agreement between [Centre] and FAO, and the recipient has no rights to obtain Intellectual Property Rights (IPR) on the germplasm or related information.

The recipient may reproduce the seed and use the material for agricultural research and breeding purposes and may distribute it to other parties provided the recipient is also willing to accept the conditions of this agreement.[1]

---

[1] This does not prevent the recipient from releasing or reproducing the seed for purposes of making it directly available to farmers or consumers for cultivation, provided that the other conditions set out in the MTA are complied with.

The recipient, therefore, hereby agrees not to claim ownership over the germplasm to be received, nor to seek IPR over that germplasm or related information. He/She further agrees to ensure that any subsequent person or institution to whom he/she may make samples of the germplasm available, is bound by the same provision and undertakes to pass on the same obligations to future recipients of the germplasm.

[Centre] makes no warranties as to the safety or title of the material, nor as to the accuracy or correctness of any passport or other data provided with the material. Neither does it make any warranties as to the quality, availability, or purity (genetic or mechanical) of the material being furnished. The phytosanitary condition of the material is warranted only as described in the attached phytosanitary certificate. The recipient assumes full responsibility for complying with the recipient nation's quarantine/biosafety regulations and rules as to import or release of genetic material.

Upon request, [Centre] will furnish information that may be available in addition to whatever is furnished with the seed. Recipients are requested to furnish [Centre] performance data collected during evaluations.

The material is supplied expressly conditional on acceptance of the terms of this agreement. The recipient's acceptance of the material constitutes acceptance of the terms of this Agreement.